Shell Shapes

by Fawn Bailey

Seashells are solids you find at the beach. The liquid water washes them up to the shore. Shells come in many strange shapes. Sometimes their shapes make them look like other objects!

You can see how this lion's paw shell got its name. It's shaped just like a paw! The shell is rough and has many ridges, but a real lion's paw is soft and furry.

Could you wear this helmet shell on your head?
It's shaped just like a real helmet, but it's too small
to wear!

Would you comb your hair with this shell? This shell is called a Venus comb. The shell has many pointy spikes, just like the combs people use. The shell isn't made of plastic, though!

Check out this moon shell! It is round and pale, just like the moon. Some moon shells are so small, you could fit them in your pocket. The moon you see in the sky is much, much bigger.

This cone shell has a familiar shape. It looks just like an ice cream cone! The shell may look like a cold treat, but it won't melt in the hot sun.

This shell is shaped like a wiggly worm. That's why it's called a worm shell. The shell may look like a worm, but it feels very different. The shell has a hard and bumpy texture. A real worm is soft and slimy.

Do you think this shell could crawl away and spin silky webs? It is called a spider conch. The shell looks like it has many legs, but it's not fuzzy like some real spiders.

This pretty shell is called a strawberry top. The shell and the berry are both red with tiny bumps. Don't bite the hard shell! Only a real strawberry is squishy and sweet.

It's easy to see how some shells get their names. Look at the shapes of these shells. What other objects do they look like? What names would you give them?

a.

b.